FREAKY TRUE SCIENCE

FREAKY SPACE STORIES

BY KATIE KAWA

Gareth Stevens PUBLISHING

Please visit our website, www.garethstevens.com. For a free color catalog of all our high-quality books, call toll free 1-800-542-2595 or fax 1-877-542-2596.

Library of Congress Cataloging-in-Publication Data

Kawa, Katie, author.
Freaky space stories / Katie Kawa.
pages cm. — (Freaky true science)
Includes bibliographical references and index.
ISBN 978-1-4824-2956-5 (pbk.)
ISBN 978-1-4824-2957-2 (6 pack)
ISBN 978-1-4824-2958-9 (library binding)
1. Astronomy—Miscellanea—Juvenile literature. 2. Space sciences—Miscellanea—Juvenile literature. I. Title.
QB46.K357 2016
520—dc23
2015006313

First Edition

Published in 2016 by
Gareth Stevens Publishing
111 East 14th Street, Suite 349
New York, NY 10003

Designer: Sarah Liddell
Editor: Ryan Nagelhout

Photo credits: Cover, p. 1 (moon and Saturn used throughout book) Morphart Creation/Shutterstock.com; cover, p. 1 (astronaut) © iStockphoto.com/eddtoro; cover, p. 1 (chimpanzee) Eric Isselee/Shutterstock.com; cover, p. 1 (Jupiter) MarcelClemens/Shutterstock.com; background throughout book Shay Yacobinski/Shutterstock.com; pp. 5, 7, 9, 11, 13, 15, 17, 19, 21, 23, 25, 27, 29 (hand used throughout) Helena Ohman/Shutterstock.com; pp. 5, 7, 9, 11, 13, 15, 17, 19, 21, 23, 25, 27, 29 (texture throughout) Alex Gontar/Shutterstock.com; p. 5 Hulton Archive/Stringer/Getty Images; p. 7 Christos Georghiou/Shutterstock.com; p. 9 photo courtesy of NASA; ESA; Z. Levay and R. van der Marel, STScI; T. Hallas; and A. Mellinger; pp. 11, 26 VICTOR HABBICK VISIONS/Science Photo Library/Getty Images; p. 13 (Makemake) Ann Feild (Space Telescope Science Institute)/KillOrDie/Wikimedia Commons; p. 13 (Haumea) West/Wikimedia Commons; pp. 13 (Ceres, Pluto, and Eris), 15 Stocktrek Images/Getty Images; p. 14 Science & Society Picture Library/Contributor/SSPL/Getty Images; p. 17 ESA/Handout/Getty Images News/Getty Images; p. 18 photo courtesy of NASA, the NASA History office, and the NASA JSC Media Services Center; p. 19 photo courtesy of NASA; p. 21 photo courtesy of NASA/JPL-Caltech/Malin Space Science Systems; p. 23 Getty Images/Handout/Getty Images News/Getty Images; p. 25 Handout/Handout/Getty Images News/Getty Images; p. 27 Photo Researchers Inc/Science Source/Getty Images; p. 28 NASA/Handout/Getty Images News/Getty Images; p. 29 (main) NASA/Getty Images News/Getty Images; p. 29 (chimpanzee) Encyclopaedia Britannica/Contributor/Getty Images.

Printed in the United States of America

CPSIA compliance information: Batch #CS15GS: For further information contact Gareth Stevens, New York, New York at 1-800-542-2595.

CONTENTS

Words in the glossary appear in **bold** type the first time they are used in the text.

THE SECRETS OF SPACE

People have been looking up at the stars and thinking about the mysteries of space for thousands of years. The study of space, or astronomy, is a branch of science that's built around solving those mysteries. As astronomers have worked to figure out the secrets of the universe, they've learned some freaky things about space. Some of these things have seemed less freaky over time, but what we now think of as basic facts were once crazy ideas!

Where do shooting stars come from? Is there water on the moon or on other planets? Exactly how big is space? These are the questions astronomers study in order to answer. However, even to the smartest scientists, a lot of things about space are still unknown.

FREAKY FACTS!

The Hubble Space Telescope can take pictures of all the planets except Mercury because it's too close to the sun.

EYES TO THE SKIES

Telescopes are an important part of discovering the stories of space. The first person to use a telescope to study space was Galileo Galilei, an Italian astronomer who lived from 1564 to 1642. Telescopes have changed a lot since Galileo's time. Now, the Hubble Space Telescope is actually in space, taking pictures of planets and other space bodies. This telescope has helped modern astronomers learn much more about space.

GALILEO WAS THE FIRST PERSON TO STUDY SPACE WITH A TELESCOPE. HE LED THE WAY FOR THE ASTRONOMERS WHO ARE NOW USING THE HUBBLE SPACE TELESCOPE AND OTHER MODERN TOOLS TO LEARN MORE THAN EVER ABOUT SPACE.

COPERNICUS CHANGES EVERYTHING

Today, it's commonly known that the sun is the center of the solar system, and Earth is one of the planets that orbit it. This thinking wasn't always the case, however. For almost 2,000 years, people believed Earth was the center of the universe.

The first person to challenge this Earth-centered model of the universe was a Polish astronomer named Nicolaus Copernicus, who was born in 1473. Copernicus came up with a theory, or idea, that the universe was actually heliocentric, which means "sun-centered." He was the first to correctly state that Earth moved around the sun, instead of the sun moving around Earth.

Copernicus's theory might not seem freaky now, but it was back when he was alive. It challenged everything people believed about the universe!

FREAKY FACTS!

Copernicus lived in a time before the telescope was used to study space. However, astronomers still knew about five planets in addition to Earth that they could see in our solar system: Mercury, Venus, Mars, Jupiter, and Saturn.

DAYS AND NIGHTS

Copernicus also found a way to explain the change from day to night in his new heliocentric model. He was the first to come up with the idea that Earth rotated, or spun, on an axis. Copernicus stated that each rotation took 1 day. These rotations create the change from day to night. This was another one of Copernicus's freaky space theories that turned out to be correct!

COPERNICUS WAS BANNED BY THE CATHOLIC CHURCH FOR HIS HELIOCENTRIC THEORY. THE IDEA WAS JUST TOO FREAKY FOR MANY PEOPLE TO EVEN CONSIDER AS FACT.

WE'RE NOT ALONE!

Another astronomer who helped change the way we saw the universe was an American named Edwin Hubble. For a long time, scientists only knew about the existence of one **galaxy**—the Milky Way—which contained Earth, the sun, and the rest of our solar system. However, in the 1920s, Hubble proved that clouds of dust existed outside of the Milky Way. This was what we could see of entirely different galaxies!

Galaxies are constantly moving in space, and they often collide, or run into each other. In fact, scientists believe the Milky Way will collide with its nearest galactic neighbor, the Andromeda Galaxy, in about 5 billion years. Scientists also believe Andromeda has already collided with other galaxies, because it has stars in it that are much younger than the galaxy itself.

FREAKY FACTS!

Some scientists believe the Andromeda Galaxy is heading toward the Milky Way at a speed of 69.4 miles (111.7 km) per second!

THIS IS WHAT SCIENTISTS THINK THE MILKY WAY AND THE ANDROMEDA GALAXIES WILL LOOK LIKE WHEN THEY COLLIDE.

WHAT'S A LIGHT-YEAR?

It will take billions of years for the Andromeda and the Milky Way Galaxies to collide because they're so far away from each other. These two galaxies are more than 2 million light-years apart! Light-years are the most common way scientists measure distances in space. A light-year is the distance light is known to travel in 1 year. One light-year is equal to around 5.9 trillion miles (9.5 trillion km)!

SEARCHING FOR OTHER WORLDS

While astronomers are always looking for new galaxies, one of the most important searches in astronomy right now is for new planets. Astronomers have found evidence of planets that exist outside of our solar system, which are called exoplanets.

Astronomers find exoplanets by looking for changes in the **electromagnetic waves** produced by stars or changes in a star's position in the sky. These changes often point to the effect a planet has on a star. When scientists notice changes in these "host stars," they often find planets orbiting them.

Astronomers hope to study these new planets to learn more about our own solar system. There's even a chance one of these exoplanets could be similar enough to Earth to have life on it!

FREAKY FACTS!

Some exoplanets are called "hot Jupiters" because they're large like Jupiter, but they orbit close to their host star. This makes them very hot.

WHERE'S THE WATER?

Scientists have used the Hubble Space Telescope to find signs of water in the atmospheres of some exoplanets. Signs of water in the atmosphere affect how light is **absorbed**. Finding these signs of water is important because scientists believe water is necessary for life to exist on a planet. In December 2013, astronomers were able to measure and compare signs of water on five different exoplanets for the first time, using tools developed for this purpose.

SCIENTISTS BELIEVE EXOPLANETS ARE THE KEY TO UNLOCKING EVEN MORE OF THE SECRETS OF SPACE.

WHAT HAPPENED TO PLUTO?

Astronomers are finding new planets, but they're also changing how we **define** what a planet is. Pluto was once known as the ninth planet in our solar system. However, in 2006, scientists stated that Pluto would no longer be known as a planet. Instead, it would become part of a group of space bodies known as **dwarf planets**.

What happened to make Pluto no longer a planet? Astronomers discovered a rocky body, which they named Eris, beyond Pluto in our solar system. At first, they wanted to name it the tenth planet, but they discovered it appeared to be larger than Pluto. This led scientists to debate what it means to be a planet and to change how a planet is defined. Because of Pluto's size and location, it was no longer considered a planet.

FREAKY FACTS!

Pluto is also known as a plutoid, which is a dwarf planet located beyond Neptune in our solar system. There are four known plutoids: Pluto, Eris, Haumea, and Makemake.

A CLOSE LOOK AT CERES

Another dwarf planet, Ceres, is located in the **asteroid belt** between Mars and Jupiter. Ceres is a very important dwarf planet because large amounts of water are believed to exist on it. There are signs of ice below its surface. Also, signs of water vapor have been discovered in its atmosphere. Some scientists believe Ceres could have more water as ice and vapor than all the freshwater on Earth!

THE DWARF PLANETS

	PLANET	YEAR DISCOVERED	LOCATION	KNOWN MOONS	AVERAGE DIAMETER	ORBIT AROUND SUN
	CERES	1801	ASTEROID BELT	0	591.8 MILES (952.4 KM)	4.6 EARTH YEARS
	ERIS	2003	KUIPER BELT	1	1,445 MILES (2,326 KM)	561.4 EARTH YEARS
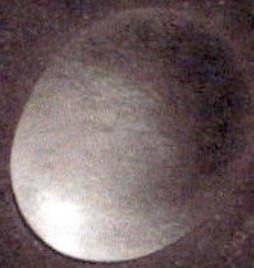	HAUMEA	2003	KUIPER BELT	2	892.3 MILES (1,436 KM)	281.9 EARTH YEARS
	PLUTO	1930	KUIPER BELT	5	1,430 MILES (2,302 KM)	247.9 EARTH YEARS
	MAKEMAKE	2005	KUIPER BELT	0	882 MILES (1,420 KM)	305.34 EARTH YEARS

ONE SMALL GROUP OF SCIENTISTS FOREVER CHANGED THE WAY WE TALK ABOUT THE SOLAR SYSTEM WHEN THEY VOTED TO MAKE PLUTO A DWARF PLANET.

PUTTING A DIFFERENT SPIN ON IT

All the planets in our solar system orbit the sun and rotate on an axis. One of those planets, Uranus, looks different from the others when it spins on its axis. It looks like it's spinning on its side as it orbits the sun. This is because its axis is tilted at a much greater angle than any other planet's axis.

Scientists believe Uranus didn't always have such a crazy rotation. They think at least two Earth-sized objects crashed into the planet at different times billions of years ago. These huge objects knocked Uranus onto its side before its moons were formed. Now those moons also seem to spin on their sides, sharing almost the same tilt to their axis as Uranus.

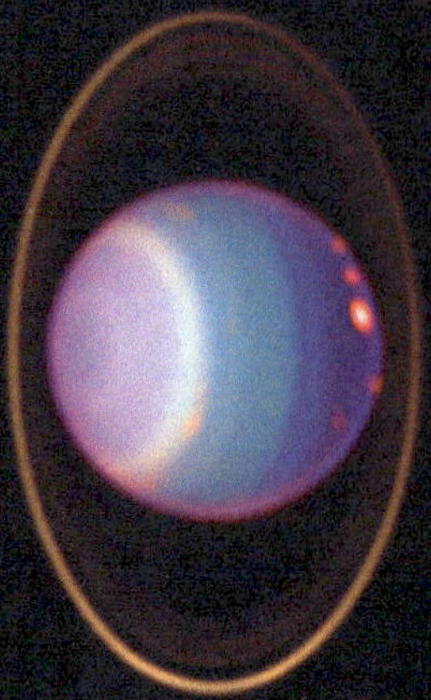

FREAKY FACTS!

Uranus has two sets of rings similar to the ones around Saturn. Uranus's rings also appear tilted because of the tilt of the planet's axis.

URANUS WAS THE FIRST PLANET TO BE FOUND USING A TELESCOPE. BRITISH ASTRONOMER WILLIAM HERSCHEL FIRST DISCOVERED THE PLANET IN 1781.

TWO DECADES OF SUMMER!

Uranus's seasons are much longer than seasons on Earth. It takes Uranus 84 Earth years to orbit the sun, so 1 year on Uranus is equal to 84 Earth years. For around 21 Earth years, the sun points directly at one of Uranus's poles because of the tilt of its axis. This means that summers on Uranus last for around 21 years! During those 21 years, however, the other side of Uranus has a very dark winter.

A CLOSER LOOK AT COMETS

Comets were one of the great mysteries of space for thousands of years. However, on November 12, 2014, a space probe, or device used to send back information about a body in space, landed on a comet for the first time. The *Philae* probe broke away from the *Rosetta* orbiter, which had been orbiting Comet 67P/Churyumov–Gerasimenko, and landed on the comet's surface.

The *Philae* probe's job is to allow scientists to better study the surface of the comet. Within a few days of its landing, it discovered traces of organic material, the building blocks of all life, on the comet's surface!

Philae's landing was a major moment in modern space exploration. It was the result of a 10-year mission for *Rosetta* to get the probe through space and to the comet.

FREAKY FACTS!

Comets have a very icy surface. *Philae* was able to land successfully and stay on the surface because it had ice screws on each of its feet, allowing it to plant itself in the ice.

A SLEEPING SPACE PROBE

The *Philae* probe was only able to send information back to scientists for 57 hours after its landing. It seems that even important space probes can have batteries that run low, and that's what happened to *Philae*. It went into a kind of sleep mode called hibernation. Scientists hope it will "wake up" when its solar panels receive more light as it gets closer to the sun.

IT TOOK 7 HOURS FOR *PHILAE* TO LAND ON THE COMET'S SURFACE!

MINING MOON WATER

Could people actually live on the moon one day? One of the first steps toward making the moon a place where people could live is getting water. But we now know that there's lots of water on the moon! In 2009, the *Lunar Reconnaissance Orbiter (LRO)* found the first signs of ice on the moon's surface. The ice is most commonly found in lunar craters and around the moon's poles.

Scientists believe this ice could be mined for drinking water and fuel for rockets. This would allow astronauts to stay on the moon for longer periods of time. The moon could even be used as a base for manned trips to Mars if people can use its water.

FREAKY FACTS!

The American astronauts who visited the moon didn't come back empty-handed. They brought back 842 pounds (382 kg) of moon rocks to Earth over 3 years.

FOOTPRINTS ON THE MOON

The moon is currently the only body outside Earth to have been visited by humans. From 1969 to 1972, the National Aeronautics and Space Administration (NASA) launched six successful missions to put humans on the moon. Twelve Americans walked on the moon, and their footprints will stay on the moon's surface for millions of years. This is because there's no wind on the moon. Some of these lunar footprints have been photographed by the *LRO* in recent years.

SCIENTISTS HOPE THE MOON WON'T BE THE ONLY SPACE BODY EVER VISITED BY HUMANS. THEY THINK MINING THE MOON FOR WATER COULD HELP MAKE HUMAN TRIPS TO MARS POSSIBLE IN THE FUTURE.

MARTIAN WATER

The moon isn't the only place in space where scientists are looking for signs of water. They've discovered signs that water could once be found on Mars, too. This means there could have been a time when Mars was able to have living things on it. This might mean there could be a time in the future when that could happen again.

How did scientists find this water on Mars? They studied the findings of NASA's *Curiosity* rover, a **vehicle** made to explore the surface of the planet. *Curiosity* found signs that Mount Sharp—a mountain in the planet's Gale Crater—was formed by **sediment** from a lake bed and rivers. According to these findings, water existed on Mars's surface for long periods of its history.

FREAKY FACTS!

Although humans have never stepped on the surface of Mars, robots have been studying the planet's surface since the *Sojourner* rover landed on Mars in 1997.

FROM SCIENCE FICTION TO REALITY

A mission to send people to Mars once sounded like the stuff of science-fiction stories. However, NASA hopes to make it a reality by the middle of the 21st century. In 2010, NASA outlined a plan to send astronauts to bodies in space where no person has ever been before. It hopes to have astronauts travel to an asteroid by 2025. Then, NASA believes they could send astronauts to Mars 5 to 10 years after that!

CURIOSITY IS HELPING NASA SCIENTISTS PREPARE TO SEND PEOPLE TO MARS BY THE 2030s!

DON'T GET SUCKED IN!

While probes are making it easier than ever before for astronomers to study things in space, there are many parts of the universe that scientists still can't directly observe. Black holes are one of these great space mysteries.

A black hole is what's left behind after a star dies. A supernova, or an explosion caused by a large star, is believed to create most black holes. After a supernova occurs, gravity causes the star to begin to collapse. The gravitational pull in the area becomes so strong that even light can't escape it.

Scientists can't directly study black holes, but they can study what they do to matter around them. If a star passes too close to a black hole, it could get ripped apart because the force of gravity is so strong!

FREAKY FACTS!

If a star or planet passes so close to a black hole that it can't escape the gravitational pull, it crosses what's called the event horizon. Once an object in space crosses the event horizon, it would need to move faster than the speed of light to escape getting sucked in to it.

ALL BLACK HOLES HAVE AN EVENT HORIZON, EVEN THE SUPERMASSIVE BLACK HOLE AT THE CENTER OF OUR GALAXY, WHICH IS SHOWN HERE.

SUPERMASSIVE BLACK HOLES

Some black holes are relatively small—only 10 to 24 times as big as the sun. Scientists believe these smaller black holes are found in large numbers throughout the universe. There could be a billion of them in the Milky Way alone! Other black holes are believed to be billions of times as big as the sun. One of these "supermassive" black holes is believed to be at the center of the Milky Way.

A NEW KIND OF ENERGY

It's often said that people slow down as they get older. When it comes to the universe, though, the opposite is true! Scientists have proof from the Hubble Space Telescope that the universe is actually expanding more quickly as time passes. When they first discovered this, it went against everything they believed about the universe.

Scientists still don't know exactly why the universe is expanding more quickly as it ages. However, they believe it's because of a mysterious energy they've named dark energy. Places scientists once thought of as empty space could contain this dark energy. As the universe gets bigger, more dark energy appears. Because there's more energy, the universe can expand more quickly. This is just one idea of the many scientists are coming up with to explain dark energy.

FREAKY FACTS!

In 2013, European scientists discovered that dark energy makes up around 68.3 percent of the whole universe!

A BATTLE OF SPACE FORCES

Dark energy is constantly battling with the force of gravity in the universe. Until scientists discovered dark energy, they believed gravity was slowing down the expansion of the universe over time. Now, they know that gravity and dark energy are working against each other. They also know that dark energy seems to be an even stronger force than gravity. If gravity were stronger than dark energy, the expansion of the universe would be slowing down—not speeding up!

SCIENTISTS STUDY DARK ENERGY BY STUDYING SUPERNOVAS SUCH AS THE ONE SHOWN HERE.

SPACE JUNK!

Space exploration is messy! People have left a lot of things behind during their time in space, and cleaning up space junk isn't as easy as cleaning up your room. This space junk that's orbiting Earth, also called orbital debris, could be a danger to spacecraft such as the International Space Station (ISS). Even the smallest piece of space junk could cause damage when it's traveling at speeds of up to 17,500 miles (28,163.5 km) per hour!

What kinds of junk can be found in space? There was a glove from Ed White, who was the first American to walk in space in 1965. There's also a camera and even a tool bag. As of 2013, NASA is tracking more than 500,000 pieces of space junk as they orbit Earth.

FREAKY FACTS!

Not all pieces of orbital debris stay in their orbit forever. Many pieces of space junk, such as White's glove, eventually burn up in Earth's atmosphere.

SHOOTING STAR OR SPACE JUNK?

In July 2014, people in Melbourne, Australia, saw a bright, fiery ball streak across the night sky. Many thought this fireball was a meteor, which is a piece of a space rock that burns up in Earth's atmosphere. However, it was actually a piece of space junk the size of a truck! The fireball was part of a Russian rocket that fell back to Earth. After wowing people with its appearance in the sky, most of the debris fell into the ocean.

NASA IS WORKING ON NEW WAYS TO TRACK ORBITAL DEBRIS AS WE SEND MORE THINGS INTO ORBIT AROUND EARTH.

SOLVING SPACE MYSTERIES

Space is so freaky because much of it remains a mystery—even to the smartest astronomers in the world. People have been studying space for thousands of years, and we've learned amazing things about the stars, planets, and other galaxies in the universe. However, there's still so much we don't know about the most distant parts of the universe and even the planets right next to Earth!

With each new discovery, scientists hope to get closer to understanding some of the freakiest and most mysterious things about space. Sending people to asteroids and Mars is one way NASA is planning to study space more closely than ever before. Imagine the freaky things we might learn from these future space missions!

FREAKY FACTS!

The ISS is heavier than the combined weight of 320 cars.

HUMANS WERE NOT THE FIRST LIVING THINGS IN SPACE. DOGS, MICE, MONKEYS, AND MANY OTHER ANIMALS HAVE BEEN SENT INTO SPACE. THE SOVIET UNION EVEN PUT A TORTOISE INTO ORBIT IN 1968.

HAM THE ASTROCHIMP

LIVING IN SPACE

Today's astronauts spend much of their time in space on board the ISS, which has been orbiting Earth since 2000 and is used for many scientific experiments. The ISS is a huge spacecraft. It's as long as a football field. It has been visited by 215 people from countries around the world. Because astronauts live on the ISS, it has to have plenty of room for them. It has two bathrooms and its own gym!

GLOSSARY

absorb: to take in

asteroid belt: the large group of small, rocky bodies in space between Mars and Jupiter

define: to describe clearly

dwarf planet: a body in space that orbits the sun and is shaped like a sphere but is not large enough to disturb other objects with its orbit

electromagnetic wave: any wave created by changes in electric and magnetic intensity, including visible light, radio waves, and X rays

galaxy: a large group of stars, planets, gas, and dust that form a unit within the universe

sediment: matter, such as stones and sand, that is carried onto land or into the water by wind, water, or land movement

vehicle: an object that moves people from one place to another

FOR MORE INFORMATION

BOOKS

Goodman, Susan. *How Do You Burp in Space? And Other Tips Every Space Tourist Needs to Know.* New York, NY: Bloomsbury Publishing, 2013.

Redfern, Nick. *True Stories of Space Exploration Conspiracies.* New York, NY: Rosen Publishing, 2015.

WEBSITES

HubbleSite Picture Album
hubblesite.org/gallery/album/
The official website of the Hubble Space Telescope features a gallery of amazing pictures taken by this telescope.

NASA Education for Students
nasa.gov/audience/forstudents/#.VL_b02TF9ng
This NASA site offers resources about a variety of space-related topics.

INDEX